STRANGE SCIENCE

ROBOTS, CYBORGS, AND AI

by Mari Bolte

Full Tilt Press
42964 Osgood Road
Fremont, CA 94539
readfulltilt.com

Full Tilt Press publications may be purchased for educational, business, or sales promotional use.

Copyright © 2024 by Full Tilt Press. All rights reserved. No part of this book may be reproduced in any manner whatsoever without written permission, except in the case of brief quotations embodied in critical articles and reviews.

All internet sites appearing in back matter were available and accurate when this book was sent to press.

ISBN: 978-1-62920-770-4 (hardcover)

ISBN: 978-1-62920-795-7 (ePUB eBook)

ISBN: 978-1-62920-799-5 (PDF eBook)

Editorial Credits

Editor: Renae Gilles

Copyeditor: Leighanna Shirey

Designer: Sara Radka

Image Credits

page 3: ©DM7 / Shutterstock; page 5: ©PhonlamaiPhoto / Getty Images; page 7: ©Dimart Graphics / Shutterstock; page 8: ©Rita Barros / Contributor / Getty Images; page 9: ©Science & Society Picture Library / Contributor / Getty Images; page 10: ©Bettmann / Contributor / Getty Images; page 13: ©kokouu / Getty Images; page 14: ©Krikkiat / Shutterstock; page 15: ©Inspiring / Shutterstock; page 16: ©SEBASTIAN KAULITZKI/ SCIENCE PHOTO LIBRARY / Getty Images; page 19: ©Capuski / Getty Images; page 20: ©Ollie Millington / Contributor / Getty Images; page 22: ©SCIEPRO / Getty Images; page 23: ©grandeduc / Getty Images; page 25: ©AXEL HEIMKEN / Contributor / Getty Images; page 26: ©PopTika / Shutterstock; page 29: ©MICHAEL BUHOLZER / Stringer / Getty Images

Cover: ©iLexx / Getty Images; ©NeoLeo / Getty Images

Printed in the United States of America.

CONTENTS

INTRODUCTION:
Beep Boop Bop 4

CHAPTER 1:
It's Alive! 6

CHAPTER 2:
Live and Learn 12

CHAPTER 3:
Revenge of the Bots 18

CONCLUSION:
Hit the Off Switch 24

Quiz 28

Activity 29

Glossary 30

Read More 31

Internet Sites 31

Index 32

Introduction

BEEP BOOP BOP

"Computer," you say, "make breakfast."

Something in the wall clicks and whirs. In a moment, there's a hot meal in front of you. You start to eat. A small robot on wheels rolls up. It is holding a tablet. You skip songs on the tablet with a flick of your eyes. The **microchip** in your brain does the rest. A robot collects your dirty dishes. It sets them in the dishwasher. The robot sees that you're low on dish soap. So it reorders some for you.

You don't notice any of that. You dress for school and jump in your hovercraft. You get seated and safely strapped in. Then the vehicle rolls out. You lean back and take a nap. There's no need to watch the road. The little ship's **artificial intelligence** (AI) knows what to do.

It sounds like something out of sci-fi. But a future among robots is closer than you think.

---

microchip: a very small electronic device

artificial intelligence: a type of computer science that makes computers seem like they have human-like thoughts and behaviors

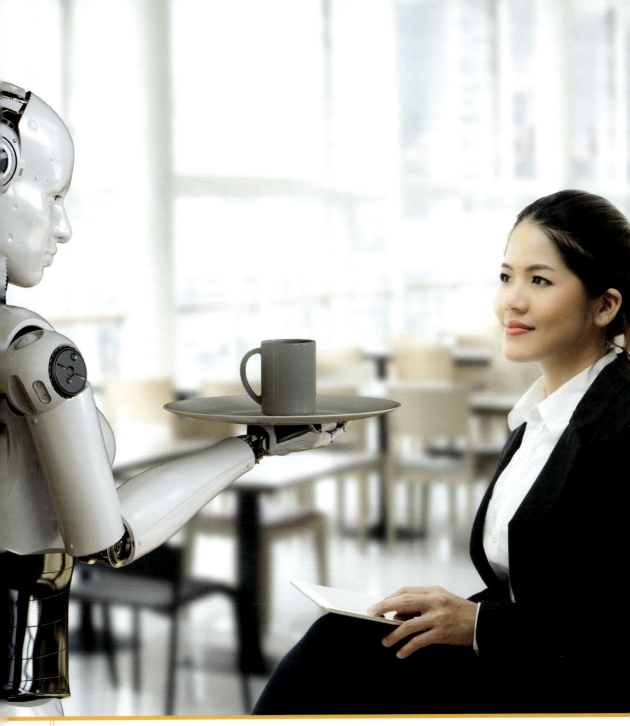

Humanoid robots integrated into everyday life are some people's dreams, and other people's nightmares.

Chapter 1
IT'S ALIVE!

Today robots are a big part of our lives. They come in many shapes. They do many different jobs. Some look like tiny bugs. Others can fly. Scientists have landed robots on Mars. There are even robots in your home. You may not have noticed them before.

When you think of robots, you might think of something with wheels or wings. But artificial beings that look like people have been part of our imagination for a long time. One Jewish tale tells of a rabbi. He lived in the 16th century. He creates a living humanlike creature out of clay. It is called a golem. The golem is meant to serve and protect. But something goes wrong. It attacks people instead of helping them. The golem must be stopped.

A golem does not have free will. Like a robot, it obeys all commands from its master.

Isaac Asimov wrote more than 500 books and short stories.

The word *robot* is somewhat new. It was in an early-1920s Czech play called *R.U.R.* The word comes from a Czech word. It means "forced labor." In the play, the robots are not made of metal. They are more like **androids**. But they look like people. And they are dangerous.

Isaac Asimov was a science professor. He wrote and edited books on advanced science. But they were written for regular people. Asimov was also a sci-fi writer. In 1940, he wrote three laws of robots. Later he added a fourth. His laws are still used today in fantasy and sci-fi stories. The laws say:

Law One: A robot may not harm humanity. By inaction, it may not allow humanity to come to harm.

Law Two: A robot may not injure human beings. Through inaction, it may not allow a human being to come to harm.

---

android: a robot that looks like a human

BACK AND FORTH

Unimate was a basic **hydraulic** moving arm. The robot weighed nearly 2,000 pounds (900 kilograms). It was easy to use in car factories. It picked things off an assembly line. Then it put them on the car's body. Those back-and-forth motions could be used for other things. In 1966, Unimate was on the *Tonight Show*. It knocked a golf ball into a cup. It poured a drink. Then it conducted the *Tonight Show* band.

Law Three: A robot must obey the orders given it by human beings, but not where such orders would conflict with the First Law.

Law Four: A robot must protect its own existence. This is as long as such protection does not conflict with the First or Second Law.

The first **industrial** robot was called Unimate. Inventor George Devol built it in 1954. In 1957, he met businessman Joseph Engelberger. Engelberger was also a physicist and engineer. He was later called the Father of Robotics. The two worked together. They developed Unimate for factory use. The robot could be used to do tasks. Its tasks were dangerous to people. Devol spoke to **investors**. He said Unimate was safe. It would follow Asimov's rules. Soon Unimates were put in factories around the country.

---

industrial: having to do with large factories

investor: a person who gives money to a company and then gets to share in the profits

Norbert Wiener first graduated from college when he was only 14 years old!

Shakey was the first robot that could make decisions. It could see where it was. The robot could also figure out how to do a task by itself. Shakey could fix its own mistakes. People could communicate with Shakey. They typed to him in English. The robot showed the world what was possible. Real-life inventions like the Roomba vacuum owe their existence to Shakey. Sci-fi characters like Rosie from *The Jetsons* do too.

Robots doing human tasks are one sci-fi dream. Mixing robots with humans is another. Norbert Wiener is called the Father of **Cybernetics**. He saw a tie between how living brains and machines work. Others saw that computers could make human bodies better. This included sci-fi writers. DC Comics' hero, Cyborg, is a mix of man and machine. His father was a brilliant scientist. Dr. Stone used experimental technology to save his son's life.

By the late 1940s, Wiener was worried. What if technology became more advanced than people? It could only lead to one winner. Machines rising up could lead to people becoming the servants. What if machines could make more of themselves? Could they be used as weapons of war? These are dangerous ideas.

> ### Strange Fact
> Wearable technology is an example of basic cybernetics. This includes smartwatches. **Implants** or external body part replacements are more advanced examples.

cybernetics: the scientific study of how people, animals, and machines control and communicate information

implant: a device that is put into the body by surgery

Chapter 2
LIVE AND LEARN

Robots are all around us. They clean our floors. They pack warehouse orders and build cars. Factory robots make sure packaged food meets standards. Household robots can set reminders. They can deliver messages. Wheeled robots follow you wherever you go. Robotic cars can drive themselves.

Our imaginations have robots doing even more work. In *Big Hero 6*, Baymax is a health-care robot. He can scan a human's body. He finds out what's wrong. Then he provides care. The droids in *Star Wars* can repair ships. They serve food and drinks. Droid armies fight real battles. Sometimes robotics become part of the human body. Darth Vader has cybernetic parts. They help him live.

In 2020, people around the world spent nearly $10 billion on robotic vacuums. That's expected to grow to $50 billion by 2028.

Star Wars characters are some of the most famous robots of all time.

AI lets robots make decisions as they do jobs for humans. It gives them autonomy. This is a being's ability to understand a situation. It makes decisions. Then it takes action. People have full autonomy. We can enter a room. We see a light bulb needs to be changed. Then we decide whether or not we will change it. We might not know how to change it. We might not know where to find new light bulbs. First we find out. Then we know for next time.

Robot autonomy is not the same in every machine. Some need a human to help them. Computers teach the robots to recognize objects. Then they learn how to follow directions. Over time, the robots may even come up with better ways to get the job done. Robots like R2-D2 and WALL-E could be the future of fully autonomous robots.

LEVELS OF AUTONOMY

In 2016, SAE International released a guide. It was about robot autonomy. It created six levels. The list was about driverless vehicles. However, it can apply to all robots.

LEVEL 0 — The robot performs basic tasks. It needs full control from a human. These are simple robots. They include washing machines and elevators.

LEVEL 1 — A human is in full control of the robot. But there are basic tools available. One example is a car with cruise control or backup assistance.

LEVEL 2 — The robot has technology that can take over basic human work. The human must be nearby to help. Robot vacuum cleaners are an example.

LEVEL 3 — The robot can do advanced human tasks. A human still needs to be ready to take over at any moment. Robots in warehouses can pick and sort orders. They move goods from one location to another.

LEVEL 4 — The robot can do most of the work itself. But it still needs a human operator. The human does not have to give their full attention. An example is aerial drones. They can fly through the air. They collect information on their own. There is still a pilot. But the drone does not need to stay where the pilot can see them.

LEVEL 5 — The robot does not need a human at all. It can learn from past experiences. It can also learn from humans and other robots. These robots can also predict events. They plan accordingly. Self-driving cars are an example.

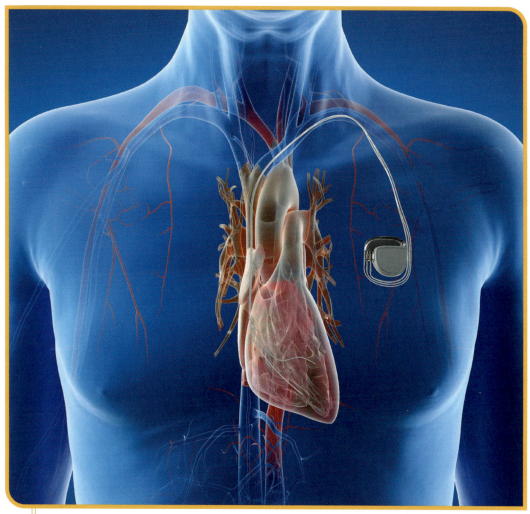

A pacemaker has a small box with a battery that connects to wire leads that go into the heart.

Cyborgs start out as people. They are already autonomous. Robotic parts just help them keep living their lives. Cybernetics seems straight out of the pages of comic books. But pacemakers are a real-life example. These are electrical devices. They can help the human heart beat at an even rate. In 1958, the first pacemaker was put into a person. Today 3 million people have pacemakers.

Biohackers take cybernetics to the next level. They implant microchips in their bodies. The chips let bodies work with machines faster than ever. These cyborgs can start a car without keys. Doors open with a wave. Meals are paid for without a wallet.

Some say biohacking is dangerous. It's experimental. How safe is putting electronics into the body? Nobody knows. There could be health risks. These range from scarring to cancer. Some biohacking cannot be undone. But biohackers say that it's their choice. The **ethics** of biohacking is based on autonomy too. Bodily autonomy is the belief that a person has the right to alter their own body as much as they want. As long as it doesn't harm anyone else, it should be allowed.

Strange Fact

Cyborgs are part human, part robot. Androids are different. They are robots that look like a person. In *Star Trek: The Next Generation*, Geordi La Forge wears a visor. It helps him to see. This makes him a cyborg. His android friend Data is a machine. He was made to look and behave like a human.

ethic: a rule that says how people should act based on ideas of what is good and bad

CHAPTER 3

REVENGE OF THE BOTS

Robots, AI, and cybernetics are moving ahead at a fast pace. Humans are welcoming them with open arms. But is technology getting too good too fast?

Roombas make the Jetsons' robot maid Rosie real. Alexa and Siri are our AI helpers at home. Cybernetic parts make everyday life easier. But are those machines helping you all the time? Listen to that soothing voice telling you about the weather. Is it really a spy? AI bots can spread misinformation online. Microchips can take over your identity. You might be trapped inside a smart house. Military robots could overthrow the government. They could even take over our nuclear weapons. With rewards come risks.

As of 2022, more than 50 percent of U.S. homes are estimated to have smart devices. This includes smart doorbells and thermostats, as well as devices like Alexa.

While *Battle Angel Alita* is from the 1990s, Japanese manga has a long and rich history that goes back hundreds of years.

AI is learning more about people every day. **Predictive AI** helps shoppers plan grocery lists. It uses **data mining**. It tracks how people make decisions. That information can be stolen. Humans could get too used to AI help. The AI could use this trust to trick people.

Iron Man has an AI butler named JARVIS. In *Avengers: Age of Ultron*, JARVIS was able to help. It stopped Ultron from getting nuclear launch codes. Later JARVIS gets an android body. It becomes Vision. Imagine an AI or android as powerful as JARVIS. What if it took sides with Ultron instead of the Avengers?

With cybernetics, humans and robots become one. The manga *Battle Angel Alita* is about a female cyborg. She makes a living as a bounty hunter. She has special robotic body parts. They let her hack computers. She can swim underwater for a long time. She can use special fighting moves. Alita looks like an ordinary human woman. Some don't realize she's more than that until it's too late.

Strange Fact

Our own bodies are working against cybernetics. They are not designed to house electronic devices. Skin irritation from implants can occur. This can cause **scar tissue** and infections. Sometimes implants even work themselves out of the body.

predictive AI: AI that uses large amounts of data from the past to make guesses about the future

data mine: to analyze

scar tissue: thick, hard cells that replace healthy ones at the site of an injury

Today one of the world's smallest robots is 0.02 inches (0.5 millimeters) wide. The largest is a humanoid vehicle. It stands at 27 feet, 9 inches (8.5 meters) tall. An evil robot of either size is a scary thought. Many working together is even scarier! Tiny robots could enter your body. They could shut down your organs. They could even destroy Earth. Billions of nanobots could malfunction. Then they could spread across the planet. They could consume everything in their path. Huge robots could flatten a city. In shows like *Power Rangers*, *Transformers*, and *Mobile Suit Gundam*, robots work together. They can merge into even larger bots.

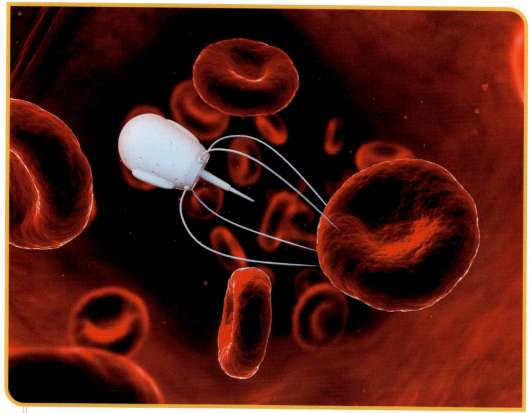

Scientists are developing nano robots that can enter the body to heal damage and disease, but what if they were to harm instead?

A ROBOT ARMY

Using robots to fight wars is often seen in sci-fi. Droid armies stand at attention in *Star Wars*. **Mecha** robots faced off in *Ready Player One*. **Sentinel** robots invaded Earth in *X-Men: Days of Future Past*. Coming face-to-face with these enemies sounds scary. They don't get tired. They don't feel pain. But wartime robots could save human lives. They would take the risks themselves. In the Terminator series, humanoid robots are assassins. They are armed with dangerous weapons. They pursue their targets across time. Robots can't be held accountable for their actions. They might not know the difference between friend or foe. They don't understand pain, hunger, or fear. How could a robot know wrong from right? There are many things to consider when it comes to a robot military.

Robots never get tired. They have many special tools. Their grips can crush you. There's a droid assassin in *The Mandalorian*. It can handle weapons with multiple arms. It can turn on the spot. No one can sneak up on it. A robot would remember being treated poorly. It could decide to take revenge.

---

mecha: a sci-fi term; having to do with giant robots
sentinel: one that stands guard or watches over

Conclusion

HIT THE OFF SWITCH

How close are these robots to reality? Military robots might be the next big thing. In 2022, nearly $19 billion was spent worldwide on research. By 2025, it might reach nearly $26 billion. Each country has its own plan. Some want to make buying and selling easier. Others want to make health care better. Still others just want cool robots.

AI can help the shopping experience. The global **economy** could be improved. In Japan, one café has robot waiters. The robots are controlled by remote. They are used by people with disabilities. The company behind the idea is Ory Laboratory. It hopes the robots can make Japan a more inclusive place. Other restaurants use robots to help out. They can take orders, bus tables, and do other simple tasks. But some worry about humans losing their jobs.

---

economy: a particular country or area's system of buying and selling things

Military developments, such as robot dogs, are leading the way in robotic research.

Being turned into a factory drone is another fear. Sci-fi writers have explored it. What if companies made employees get microchipped? Some states have taken moves to stop this from happening. But there are few laws around biohacking. And weaponizing cybernetic people or prosthetics is always a possibility.

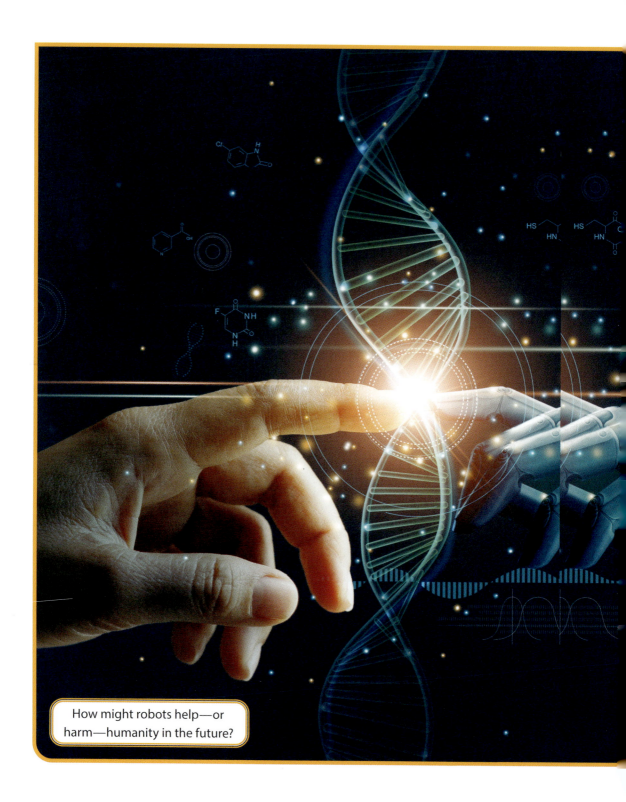

How might robots help—or harm—humanity in the future?

Nobody knows what adding more robots to the world might do. There are many questions and no answers. We only have sci-fi theories and ideas. Building robots uses a lot of materials. Then they need energy to run. They put out a lot of waste. How long would they last? What would we do with outdated robots?

Would cybernetics be available to everyone? Or just those who could afford it? Right now, there are no rules about implanting chips or parts. What if a person's microchip was used to invade their privacy?

Inventing new devices means dealing with new rules and new complications. Like AI, humans will have to learn on the go. Hopefully, we will be able to stay ahead of the machines. If not, we may find ourselves at their mercy.

QUIZ

#1
How many levels of robot autonomy are there?

a. five
b. six
c. seven
d. four

#2
The first industrial robot was called:

a. Baymax
b. Unilever
c. Unimax
d. Unimate

#3
Which statement is part of Law Four of robotics?

a. A robot must protect its own existence.
b. A robot may do advanced human tasks.
c. A robot may not have free will.
d. A robot may not harm humanity.

#4
Which can be a side effect of cybernetics?

a. technology becoming more advanced than people
b. misinformation
c. scar tissue and infections
d. data mining

1. b 2. d 3. a 4. c

ACTIVITY

The Cybathlon is also known as the Bionic Olympics. People from around the world compete in this special event. They show off their prostheses and AI assists. Events have two winners. One goes to the athlete. The other goes to the person behind the technology. Other bionic athletes around the world participate in sports like skiing, swimming, and running.

Advancements in robotics, AI, and cybernetics continue to enhance human life. Research a bionic athlete. What sport do they compete in? Find out more about their prosthetics. What are they made from? How do they work? How much control does the athlete have? Draw a diagram of the athlete's bionic part. Then share your research with another person.

GLOSSARY

android: a robot that looks like a human

artificial intelligence: a type of computer science that makes computers seem like they have human-like thoughts and behaviors

cybernetics: the scientific study of how people, animals, and machines control and communicate information

data mine: to analyze

economy: a particular country or area's system of buying and selling things

ethic: a rule that says how people should act based on ideas of what is good and bad

hydraulic: made to move by using fluids and pressure

implant: a device that is put into the body by surgery

industrial: having to do with large factories

investor: a person who gives money to a company and then gets to share in the profits

mecha: a sci-fi term; having to do with giant robots

microchip: a very small electronic device

predictive AI: AI that uses large amounts of data from the past to make guesses about the future

scar tissue: thick, hard cells that replace healthy ones at the site of an injury

sentinel: one that stands guard or watches over

READ MORE

Andrews, John. *Bots and Bods.* Kansas City, MO: Andrews McMeel Publishing, 2021.

Kachala, Elaine. *Superpower?: The Wearable-Tech Revolution.* Custer, WA: Orca Book Publishers, 2022.

Mattern, Joanne. *All about Artificial Intelligence.* Cutting-Edge Technology. Lake Elmo, MN: Focus Readers, 2023.

INTERNET SITES

https://www.aventine.org/robotics/history-of-robotics
13 Milestones in the History of Robotics

https://academickids.com/encyclopedia/index.php/Isaac_Asimov
Academic Kids Encyclopedia: Isaac Asimov

https://kids.kiddle.co/Cyborg
Cyborg Facts for Kids

INDEX

androids 8, 17, 21
Asimov, Isaac 8, 9
autonomy 14, 15, 17

biohacking 17, 25

DC Comics 11
Devol, George 9

factories 9, 12, 25

golems 6–7

health care 12, 24

mecha robots 23
microchips 4, 17, 18, 25, 27
military robots 18, 24–25

pacemakers 16

robot vacuums 11, 12–13, 15
R.U.R. 8

self-driving cars 15
Shakey 11

Unimate 9

Wiener, Norbert 10–11